CW00641074

THE ILLUSTRATED MOTORCAR LEGENDS

PORSCHE

ROY BACON

SUNBURST BOOKS

Acknowledgements

The author and publishers wish to acknowledge their debt to all those who loaned material and photographs for this book. The bulk of pictures came from the extensive archives of the National Motor Museum at Beaulieu. We also had kind assistance from Porsche Cars of Great Britain, and one picture was from Sotherby's, the auctioneers. Thanks to all who helped.

Published 1995 by Sunburst Books, Deacon House, 65 Old Church Street, London SW3 5BS.

ISBN 1 85778 145 7
Printed and bound in China

CONTENTS

FOUNDATIONS

An early pre-war step in the evolution of the Volkswagen, one of the many Dr Ferdinand Porsche designs that preceded his own marque.

Porsche – teutonic but light, squat and stylish, always impressively fast, the car of the upwardly mobile of the 1980s, coveted by many and owned by those who enjoy their driving – a car with individuality and originality.

Dr. Ferdinand Porsche designed cars for many firms before establishing his own. In Edwardian days there was the Lohner-Porsche which used its engine to drive an electric generator to power motors in its road wheels. Then came Austro-Daimler, Mercedez, Steyr, the foundation of the Porsche design studio in 1930, Wanderer, the fabulous V-16 Auto Union racing car of the 1930s and, at Hitler's behest, what became the Volkswagen. By then, his son, Ferry, had joined the firm and fresh designs flowed from it for all kinds of items.

Postwar, Ferry and his sister kept the firm alive while their father remained in prison. A major project to design a grand prix car for Cisitalia provided the funds to secure Ferdinand's release and gave the firm a sound foundation, but by then, he was both old and in ill-health. He died in 1951, but happy in the knowledge that cars bearing his name were already in production and the firm secure in the hands of his son.

The Cisitalia grand prix car which funded Ferdinand's release and the start of the firm. An exotic design using a flat-12, supercharged, 1.5-litre engine, it gave Porsche invaluable data which was used in later years.

356: STOCK AND SPYDER

Porsche allocated job numbers to their design projects in sequence and thus, 356 became the one which was to establish their name as a car manufacturer. Thanks to their Volkswagen link, Porsche were able to use many VW parts to create their own and the first was built in 1948 at their establishment at Gmünd, in Austria, to which they had moved during the war to avoid the bombing.

That first Porsche was a mid-engined car, its tuned, 1131cc Volkswagen flat-four engine mounted ahead of the rear axle which it drove via a four-speed gearbox. The mechanics went into a tubular space frame, there was all-round independent suspension, and the open body was hand-built in aluminium. As with all Porsche cars, it was low, sleek and stylish, while the engine was tuned and many of the VW parts modified in some way for an improvement.

The 356/2 of 1949, the first production Porsche, built at Gmünd in Austria before the firm moved back to Stuttgart.

Above: Centre fold from the 1950 Porsche brochure which shows the platform chassis, general layout of the car, the dash and an engine section. All served the firm well for many years.
Below: Early Gmünd-built 356/2 with its upright windscreen wipers, quarter lights, split screen and semaphore trafficators.

Another page from the 1950 brochure showing the Stuttgart-built coupe. Alongside are shown the cars Dr Porsche was involved with, from 1900 on.

Later in 1948 the first true production cars were built as the 356/2. They differed a good deal from the first for the engine was mounted behind the rear axle and the space frame changed to a welded sheet-steel platform with box sections and a central tunnel to give it stiffness.

The engine continued to be heavily based on the VW unit, often with the bore reduced to bring the capacity down to 1086cc to suit the 1100cc competition class. The flat or opposed boxer four-cylinder layout remained in use, along with the air cooling of the VW and its single, central camshaft opening its overhead valves via pushrods and rockers. Breathing through two Solex carburettors, it produced some 40 bhp at a low 4,200 rpm, when the VW made only 25.

The gearbox came from VW and lacked synchromesh but suspension remained independent all round. At the front went parallel trailing arms and at the rear a swing axle, all controlled by torsion bars and hydraulic shock absorbers. Brakes were drum, mechanically cable operated as for the VW, and the steel, bolt-on wheels carried 5.00 x 16 inch tyres.

The two-door, coupe body was hand-built in aluminium, characterised by its fine shape, air vents in the engine cover and shallow rear window. The windscreen was a split-vee type and the early cars had front quarter lights. The wiper blades parked in the vertical and semaphore trafficators were fitted rather than turn signals. There was little exterior trim and the inside was plain. The rev-counter of the first car no longer fitted beside the speedometer.

Inevitably, production was limited, body building the problem. Thus, around 50 cars were built at Gmünd, mainly coupes but with eight cabriolets, six bodied by Beutler in Switzerland and the others by Porsche. The family sought to overcome this and to return to Stuttgart, succeeding in both aims early in 1950. They contracted the body building to Reutter and in April 1950 the first German-built Porsche was produced.

A 356 with a cabriolet body on show at Earls Court for the London motor show, one of the first to have right-hand drive.

The move to Stuttgart brought other changes: the cars adopted the hydraulic brakes introduced on the VW and sales rose dramatically. In part this was due to the dynamic lines and good performance, but was also enhanced by the start of the Porsche involvement in competition. While the stock 356 was first raced in the USA, in Europe Walter Glöckler became involved with the factory.

Following a single-seater built in 1950 came sports-racing cars known as Glöckler-Porsches fitted with tuned Porsche engines under open aluminium Spyder bodies. These were immediately successful, taking class wins at Le Mans from 1951 to 1958, and winning many other and varied events.

In 1951 the production 356 changed to twin-leading-shoe front brakes and lever-arm rear shock absorbers which replaced the telescopic type to match those at the front. The original engine, known as the 1100, was joined by a 1300, its actual 1287cc capacity achieved by boring out the earlier one to produce 46 bhp at the same engine speed. Its specification included Mahle aluminium cylinders.

Late in 1951, these two were joined by the 1500 which kept the 1300 bore but lengthened the stroke to arrive at 1488cc. It introduced the roller-bearing crankshaft, but not for the usual reasons of less friction and a higher engine speed. The design allowed for one-piece connecting rods whose smaller dimensions enabled the longer stroke to swing in the existing crankcase, not for their usual reason of improved strength. Produced by Hirth, the crankshaft was an assembly of parts located to each other by serrations and clamped using differentially-threaded screws – very rigid, special and expensive.

Two early Porsches, from 1949 and 1951: the enclosed wheels were typical of the cars run in sports car races of that era. Early successes included class wins at Le Mans.

Changes followed thick and fast, inevitable for a car based on a low-powered stock vehicle, intended for fast driving, and often used in competition. A useful increase in brake drum diameter and width came in 1952 along with a Porsche synchromesh gearbox, some body alterations and a folding rear seat. A cabriolet body joined the coupe to offer open-air motoring while a lightweight roadster was built in small numbers at the behest of the USA distributor to become the America. Most went to the USA where they were raced by private owners, and to suit this, the bodies were in aluminium and the windscreen detachable.

By 1953 a plain-bearing crankshaft 1500 was available and the roller version became the 1500S. It went into the two 550 models built by Glöckler that year to race in European events where they had considerable success. This led, in 1954, to a production-racing Porsche which was listed as the 550/1500RS or 550 Spyder, the RS indicating 'rennsport', German for 'racing sport'. It was fitted with a new 1498cc dry-sump engine featuring twin overhead camshafts driven by shafts and gears. These units were known as four-cam engines and had twin plugs for each pair of cylinders and a Hirth built-up crankshaft with roller bearings. The car had a tubular chassis under a light, aluminium open body which came with a convertible top for around twice the price of a stock 356, itself hardly a cheap car. About 80 of these cars were built, mainly in 1955 and they chalked up many successes.

Line up of racing Porsches in 1953, at Le Mans to run in the 1500cc class, which they won. Fitted with the 1498cc four-cam engine with shaft and gear-driven twin overhead camshafts on each bank and a Hirth built-up crankshaft.

Meanwhile, the 356 continued in 1953 with 1100, 1300 , 1500 or 1500S engines under a coupe or cabriolet body. Sales in the USA struggled due to the high prices, so in 1955 the importer suggested and the factory produced the 1500 America which became known as the Speedster. The cabriolet body was used, stripped of many features which were standard in the 1500 Super. This pulled the price under the $3,000 mark, although the customer had to pay extra for a heater.

During 1954 the 1100 model was dropped but a 1300S with a roller-bearing crankshaft was added although it was short-lived. The engines were reduced to four for 1955, the 1300 and 1500 with plain-bearing crankshaft and the 1300S and 1500S with roller assemblies. All went into a new three-piece crankcase which replaced the older two-piece magnesium one. It was a major step forward and a move away from being dependent on VW. An anti-roll bar was added to the front suspension whose spring rate was altered and this took the original 356 up to the end of the year when a revised series was announced for 1956.

PORSCHE
UUW 14

Above: The convertible body which gave open-air motoring and was equally popular with Porsche enthusiasts.

Opposite: Early 356 displaying the lovely lines of the marque and type which set it off so well, with conventional wipers and turn signals.

49
49

ABOVE: Porsche 550 Spyder, built for racing with a high-tech specification for its time. Quick and very successful with fine lines.

OPPOSITE TOP: Interior of the convertible 356 which was well set out. The rear seats could fold down to increase the storage capacity.

OPPOSITE BOTTOM: The works RSK team in 1954. Much of the technology learned running these cars went into the RS or rennsport models sold to private owners for sports car racing. Some of these became super road cars.

RIGHT: The four-cam engine which was based on the stock flat-four but had twin overhead camshafts. The vertical tube between these encloses one of the drive shafts. A complex, expensive but very powerful engine.

BELOW: The 1954 brochure which gave some background and history to the marque as well as featuring the 356.

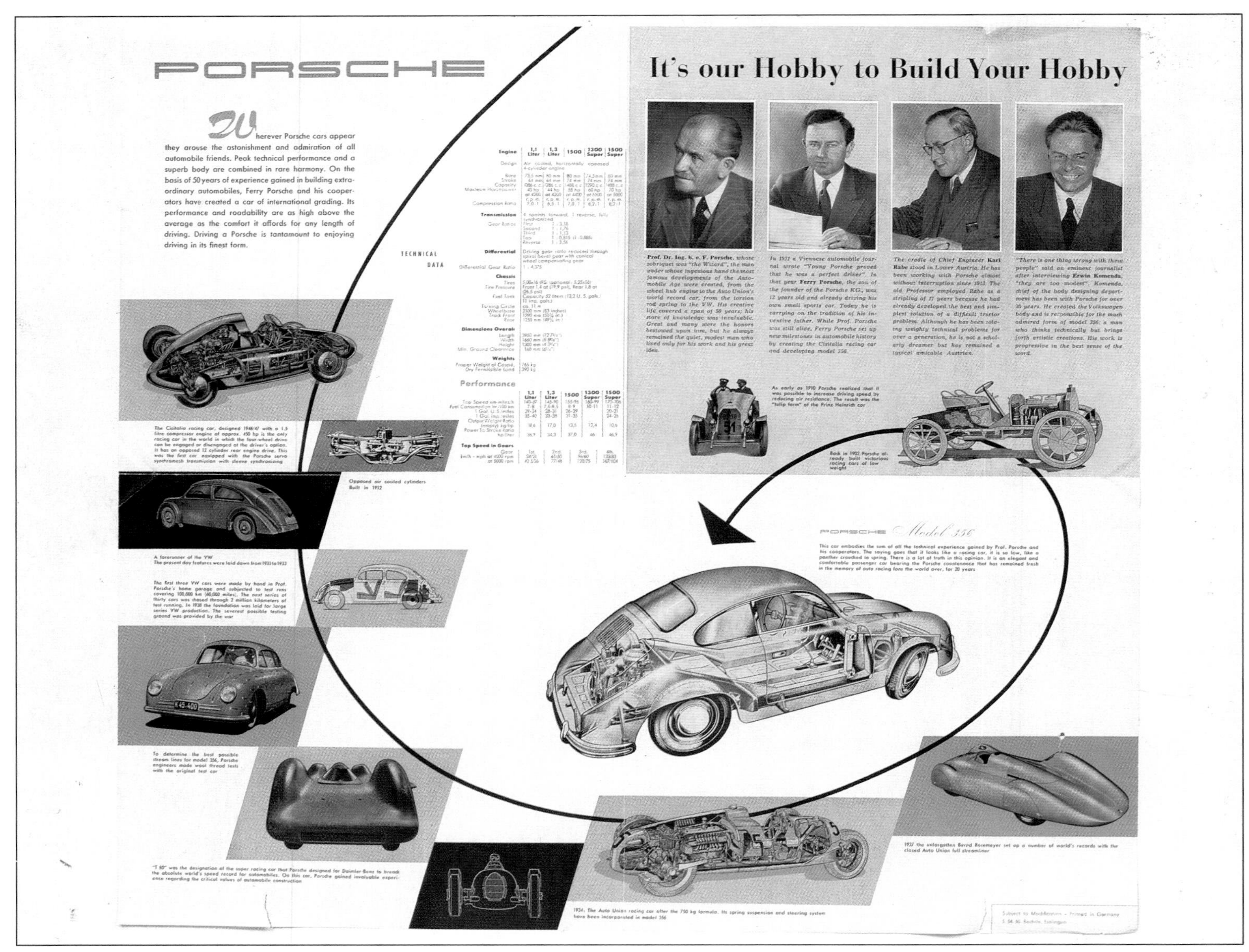

PORSCHE

Wherever Porsche cars appear they arouse the astonishment and admiration of all automobile friends. Peak technical performance and a superb body are combined in rare harmony. On the basis of 50 years of experience gained in building extraordinary automobiles, Ferry Porsche and his cooperators have created a car of international grading. Its performance and roadability are as high above the average as the comfort it affords for any length of driving. Driving a Porsche is tantamount to enjoying driving in its finest form.

TECHNICAL DATA

Engine	1,1 Liter	1,3 Liter	1500	1300 Super	1500 Super
Design	Air cooled, horizontally opposed 4-cylinder engine				
Bore	73,5 mm	80 mm	80 mm	74,5 mm	80 mm
Stroke	64 mm	64 mm	74 mm	74 mm	74 mm
Capacity	1086 c.c.	1286 c.c.	1488 c.c.	1290 c.c.	1488 c.c.
Maximum Horsepower	40 hp at 4200 r.p.m.	44 hp at 4200 r.p.m.	55 hp at 4400 r.p.m.	60 hp at 5500 r.p.m.	70 hp at 5000 r.p.m.
Compression Ratio	7,0 : 1	6,5 : 1	7,0 : 1	8,2 : 1	8,2 : 1

Transmission 4 speeds forward, 1 reverse, fully synchronized
Gear Ratios: First 1 : 3,18; Second 1 : 1,76; Third 1 : 1,13; Top 1 : 0,815 (1 : 0,885); Reverse 1 : 3,56

Differential Driving gear ratio reduced through spiral bevel gear with conical wheel compensating gear
Differential Gear Ratio 1 : 4,375

Chassis
Tires 5,00x16 (PS: optional: 5,25x16)
Tire Pressure Front 1,4 at (19,9 psi), Rear 1,8 at (26,5 psi)
Fuel Tank Capacity 52 liters (13,2 U. S. gals./11 imp. gals.)
Turning Circle ca. 11 m
Wheelbase 2100 mm (83 inches)
Track Front 1290 mm (50¾ in.)
Rear 1250 mm (49¼ in.)

Dimensions Overall
Length 3950 mm (12.7½")
Width 1660 mm (5.5⅜")
Height 1300 mm (4.3¼")
Min. Ground Clearance 160 mm (6¼")

Weights
Proper Weight of Coupé 765 kg
Dry Permissible Load 390 kg

Performance

	1,1 Liter	1,3 Liter	1500	1300 Super	1500 Super
Top Speed km-miles/h	140-87	145-90	155-96	160-99	170-106
Fuel Consumption ltr./100 km	7-8	7,5-8,5	8-9	10-11	11-12
1 Gal. U. S./miles	29-34	28-31	26-29		20-21
1 Gal. imp./miles	35-40	33-38	31-35		24-25
Output/Weight Ratio (empty) kg/hp	18,6	17,0	13,5	12,4	10,6
Power To Stroke Ratio hp/liter	36,9	34,3	37,0	46	46,9

Top Speed in Gears

Gear	1st.	2nd.	3rd.	4th.
km/h - mph at 4000 rpm	34/21	61/30	96/60	133/83
at 5000 rpm	42.5/26	77/48	120/75	167/104

The Cisitalia racing car, designed 1946/47 with a 1.5 litre compressor engine of approx. 450 hp is the only racing car in the world in which the four-wheel drive can be engaged or disengaged at the driver's option. It has an opposed 12 cylinder rear engine drive. This was the first car equipped with the Porsche servo synchromesh transmission with sleeve synchronizing

Opposed air cooled cylinders Built in 1912

A forerunner of the VW
The present day features were laid down from 1931 to 1932

The first three VW cars were made by hand in Prof. Porsche's home garage and subjected to test runs covering 100,000 km (60,000 miles). The next series of thirty cars was chased through 2 million kilometers of test running. In 1938 the foundation was laid for large series VW production. The severest possible testing ground was provided by the war

To determine the best possible dream lines for model 356, Porsche engineers made wool thread tests with the original test car

"T 80" was the designation of the super racing car that Porsche designed for Daimler-Benz to break the absolute world's speed record for automobiles. On this car, Porsche gained invaluable experience regarding the critical values of automobile construction

It's our Hobby to Build Your Hobby

Prof. Dr. Ing. h. c. F. Porsche, *whose sobriquet was "the Wizard", the man under whose ingenious hand the most famous developments of the Automobile Age were created, from the wheel hub engine to the Auto Union's world record car, from the torsion rod spring to the VW. His creative life covered a span of 50 years; his store of knowledge was invaluable. Great and many were the honors bestowed upon him, but he always remained the quiet, modest man who lived only for his work and his great idea.*

In 1921 a Viennese automobile journal wrote "Young Porsche proved that he was a perfect driver". In that year **Ferry Porsche**, *the son of the founder of the Porsche KG., was 12 years old and already driving his own small sports car. Today he is carrying on the tradition of his inventive father. While Prof. Porsche was still alive, Ferry Porsche set up new milestones in automobile history by creating the Cisitalia racing car and developing model 356.*

The cradle of Chief Engineer **Karl Rabe** *stood in Lower Austria. He has been working with Porsche almost without interruption since 1913. The old Professor employed Rabe as a stripling of 17 years because he had already developed the best and simplest solution of a difficult tractor problem. Although he has been solving weighty technical problems for over a generation, he is not a scholarly dreamer but has remained a typical amicable Austrian.*

"There is one thing wrong with these people" said an eminent journalist after interviewing **Erwin Komenda**, *"they are too modest". Komenda, chief of the body designing department has been with Porsche for over 20 years. He created the Volkswagen body and is responsible for the much admired form of model 356: a man who thinks technically but brings forth artistic creations. His work is progressive in the best sense of the word.*

As early as 1910 Porsche realized that it was possible to increase driving speed by reducing air resistance. The result was the "tulip form" of the Prinz Heinrich car

Back in 1902 Porsche already built victorious racing cars at low weight

PORSCHE *Model 356*

This car embodies the sum of all the technical experience gained by Prof. Porsche and his cooperators. The saying goes that if it looks like a racing car, it is so low, like a panther crouched to spring. There is a lot of truth in this opinion. It is an elegant and comfortable passenger car bearing the Porsche countenance that has remained fresh in the memory of auto racing fans the world over, for 20 years

1937 the unforgotten Bernd Rosemeyer set up a number of world's records with the closed Auto Union full streamliner

1934: The Auto Union racing car after the 750 kg formula. Its spring suspension and steering system have been incorporated in model 356.

Subject to Modifications – Printed in Germany

Speedster on show, a body style built at first for the USA but popular all over.

Above: A Porsche 1100 running in the 1954 Monte Carlo Rally, not quite the event associated with the marque but it coped well enough.

Right: Fine 356 Speedster with the top raised to combat the British weather.

473 FMT
PORSCHE

356: A, B AND C

Porsche unveiled the 356A range at the 1955 Frankfurt show to continue much the same body line. On the outside there was a one-piece, curved windscreen, inside went a padded dash and reclining seats plus improved instrumentation.

For general use, two engine sizes were offered, the 1300 and the 1600 which had a larger bore. Both also came in the S version with a roller-bearing crankshaft, and all remained in the flat-four layout with pushrod-operated, overhead valves. The gearbox remained a Porsche-built, all-indirect, four-speed synchromesh type. The suspension was softened and the wheel size reduced to 15 inches, but widened. The Coupe, Cabriolet and Speedster bodies continued to come from Reutter, and the overall result was an improved car, in the Porsche style. Which meant quick, tail-heavy although less so than for the 356, limited luggage space and some wind noise, but one of the most satisfying cars in the world to drive. Responsive, rewarding, refreshing – a car the driver wore rather than simply sat in.

For 1956 Porsche produced the 356A model and this is the coupe fitted with the 1600 Super engine as offered for 1958. The lines are little changed but the car was improved in many ways.

There were detail changes to both the bodies and engines for 1957, during which year the gearbox housing was changed. From being split on the centre line it became a tunnel type into which the gears and shafts went from one end. The 1300 engine was dropped for 1958 while the 1600S changed to a plain-bearing crankshaft. Following factory approval of a detachable hardtop marketed by a US firm a year or two earlier, a Porsche version was listed for the cabriolet, its finish colour usually contrasting with that of the body. The Speedster had a revised body built by Drauz for 1959 to become the Speedster D at first, and later the Convertible D, while the Reutter-built Coupe and Cabriolet models continued.

In September 1959 Porsche unveiled the 356B models at the Frankfurt show to introduce a new body style. Both the headlights and bumpers were raised to give a new line, while detail changes emphasised this. The convertible continued to be built by Drauz, but was renamed the Roadster, while Reutter produced the coupes and cabriolets. Some mechanical changes enabled the rear seats to be lowered for more headroom although the cars remained cramped for four. There were new brake drums, cast in aluminium with radial cooling fins and the iron liners bonded in place. The engines were little altered from the earlier cars, being the 60 bhp 1600 and 75 bhp 1600S, both of 1582cc.

In March 1960 these two were joined by the Super 90 or 1600S-90, modified to produce 90 bhp and given a refined lubrication system to cope with the extra load and high cornering speed. Many of the detail changes filtered onto the other engines in time.

Cabriolet body on a 1958 1600 356A, for which a detachable hardtop was also available.

Above: By 1960 the model had become the 356B, and this is the coupe version which continued the lines as of old, but with the bumpers and headlights raised.

Opposite top: Hardtop type 356B for 1960, a fine car built essentially for two to travel fast in, even if four could be crammed in for a short ride.

Opposite bottom: Sprint version of the Speedster which has lost some weight thanks to the removal of the bumpers and hub caps. A 1959 car.

There were suspension improvements for 1961 when some Roadster bodies were built by D'Ieteren of Brussels. In addition, Karmann began producing a Hardtop Coupe, taking over some Coupe production from 1962 on, mostly to assist Reutter and increase Porsche build numbers. The bodies were revised again for 1962 along with much of the mechanics, continual development being the never-ceasing theme of Porsche.

During 1963 Porsche absorbed the Reutter body firm and in July began to build the final version of the series, the 356C. This kept the same basic bodies. Reutter built both coupe and cabriolet, Karmann just the coupe, while the most important change was to disc brakes on all four wheels. Porsche had been working on discs since 1958, developing their own light system, but for production settled for a Dunlop design, made by Teves, but adapted to suit Porsche and incorporate their ideas.

Just two 1582cc engines were listed for the standard road cars, the 1600C and 1600SC, both essentially as the S and S-90 already in use but further refined. In this form the 356 ran on to the end of 1965 in the USA but was replaced in Europe during the year. It brought the total built to over 76,000, the 356B the most prolific at close to 31,000. A handful, ten more, were built in 1966 for special friends.

ABOVE: The Super 90 or 1600S-90 as built for 1960 to give more power and performance, in this case with the cabriolet body.

OPPOSITE TOP: Cockpit and dash of the 1960 Super 90 which had clear dials right in front of the driver where they should be, clock in the centre for the passenger and minor dials to the right.

OPPOSITE BOTTOM: The 356C as shown in the mid-1963 brochure. It kept the same body styles but moved to disc brakes on all wheels, all models fitting the 1600 engine in S or S-90 forms.

PORSCHE

CARRERA AND SPYDER

A Carrera 356 introduced for 1956 as the 1500 GS and fitted with the 1498cc four-cam engine in a detuned form. Striking, fast and listed in all three body styles, this is the coupe.

When Porsche introduced the 356A at Frankfurt late in 1955, a striking new model stood alongside the road cars. This was the Carrera 1500GS, named after the Mexican race in which Porsche had been most successful. It was powered by a detuned version of the 1498cc, four-cam 550 Spyder so kept the twin overhead camshafts, Hirth roller-bearing crankshaft, dry-sump lubrication, twin ignition coils and two plugs per cylinder. It delivered 100 bhp at 6,200 rpm.

The Carrera was built in all three body styles, distinguished by gold script markings and ran best between 2,500 and 6,500 rpm. It could go to 7,500 with safety but this drank petrol and increased wear, while low-speed lugging was not kind to the bearings.

The Spyder became the type 550A/1500RS for 1956, and had a new and stiffer tubular frame, revised rear suspension and less weight. The works car won the Targa Florio, a wonderful demonstration of its abilities, and other events so that late in 1956 Porsche acceded to demands and built a batch of cars for sale. These had bodies by Wendler and a convertible top as required by the 1957 rules. In all 37 were built during 1956 and 1957 to be raced with many successes.

The Porsche Carrera on show around 1957, attracting much interest.

By May 1957 there were two versions of the Carrera. The De Luxe became the true road car, as it had more home comforts together with the Porsche style and performance. For racing, the model was the Gran Turismo (GT) which was built in both Coupe and Speedster forms but lost weight, had the power raised to 110 bhp at 6,400 rpm, alternative gearbox and axle ratios, but no comforts.

While the production 550A continued, the works moved on to the RSK during 1957, which had redesigned front suspension that year and a Watts linkage at the rear for 1958. Engines were either of 1498 or 1587cc while many new ideas were tried out at various events.

Some of this arose due to Formula 2 racing being for 1500cc cars for which the Spyders were eligible. This led to an RSK being adapted to a centre seat in 1958 and others having the choice of centre or offset for 1959. By then it was known that for 1961 the Formula 1 limit would be 1500cc and this was to affect Porsche thinking.

In 1958 the Carrera GT lost weight thanks to aluminium doors, engine cover, front lid and bucket seats, plus other measures. The engine was revised internally to allow for plain bearings but kept to rollers and 1.5-litre for 20 units. Then came 14 of the same size but having a plain-bearing crankshaft while a 1.6-litre version was on trial for use in 1959. The GT continued to be successful but the De Luxe was not the strong seller that had been hoped for. Too special for general road use and not suitable for competition, it fell between the two stools.

An early Formula 1 effort at Monaco in 1959, with Wolfgang von Trips at the wheel of the 1500cc car.

The GT continued for 1959, usually fitted with the four-cam engine although a few went to California fitted with the 1.6-litre ohv unit. The De Luxe was revised into the Carrera 1600GS which was the most luxurious Porsche built up to that time. The engine was the stretched out, plain-bearing crankshaft type of 1587cc as used in the GT but detuned a little. The fixtures and fittings were de luxe and these, plus more sound insulation, put the weight up so the performance was no better and had lost the snap of the past. While more comfortable than before, this was not the attribute at the top of the Porsche buyer's list. The result was fewer than 100 examples sold during the model's short, two-year life.

Meanwhile, in 1959 came the first production run of the RSK models and a factory car which met the proposed Formula 1 rules, with exposed wheels. It had a six-speed gearbox and ran in Formula 2 that year while the Spyder 1500RSK cars were found to be very taut, their handling on a knife edge so hard to control, but quick for those drivers who managed.

Both the Carrera and Spyder changed in 1960, a year when Porsche laid plans to compete in Grand Prix racing. To achieve this, their first step was to update the RSK to become the RS60 which had a wider frame to suit the sports car race regulations, better suspension, longer wheelbase and either 1498, 1587cc or 1679cc engines. The car still came with a convertible top (the rules called for this). Four were built for the works team and a dozen for private runners. This brought more success in Formula 2 while work was begun on an eight-cylinder Grand Prix engine.

From the RSK, Porsche developed this RS60 for sports car racing, with the screen and hood as required by the rules. As well as the works cars, some were sold to private owners.

The Carrera continued in lightweight form only. Reutter was asked to build GT coupe bodies which went over the 1.6-litre, four-cam engine producing 115 bhp at 6,500 rpm and, for the first time, 12-volt electrics were used. More special were the cars fitted with all-aluminium bodies by Zagato of Italy. This was arranged through Carlo Abarth and the cars were much lower, narrower, minus bumpers and without frills. Known as the Abarth Carrera GTL, only about 20 were built to be one of the most exotic and exciting of all the road Porsche cars.

In 1961 the Carrera continued with a stronger bottom end for the engine and a power output that varied according to whether the stock or the open-pipe Sebring exhaust was fitted. The firm entered Grand Prix racing using fuel injection but with little success. One car was run with its cooling fan laid flat above the engine and disc brakes but the factory reverted to using Formula 2 cars having carburettors for most races. On the sports-racing side the 1960 car was re-typed the RS61 and joined by the W-RS, an open-topped Spyder fitted with a 1966cc, four-cam engine.

Late in 1961 the Carrera 2000GS made its debut at Frankfurt. Listed as the 356B/2000GS and sold as the Carrera 2, it had a fully-trimmed Reutter de luxe coupe or cabriolet body, Porsche disc brakes and a 1968cc engine. The extra capacity came from a longer stroke and a forged crankshaft with plain bearings was fitted, along with a bigger oil pump. Over 400 Carrera 2s were built, the last 120 or so based on the 356C. The type was homologated with the Abarth body and a tuned engine in order to allow the GTL to use the 2-litres and disc brakes.

Above: Fabulous Abarth Carrera GTL, bodied by Zagato and perhaps the most desirable of all the early Porsche cars. Only 20 of these exciting machines were built.

Opposite: Porsches at Monaco in 1961, Bonnier's car is nearest but failed to finish while Gurney's older Formula 2 car was fifth.

On the Grand Prix front some Formula 1 cars were sold off while the flat-eight engine was fitted into a lengthened chassis. It continued the known features of twin overhead camshafts driven by shafts and gears, won the 1962 French Grand Prix at Rouen, was reliable, but was not raced in 1963. Porsche had more success with the W-RS which ran with a 2-litre version of the flat-eight engine in 1962 and went on to win the 1963 Targa Florio and the hill climb championship that year and in 1964.

In 1963 a new Carrera body appeared on a RS61. It had an aggressive line and minimal trim to be raced by the works who moved on to using the Carrera GTS, also known as the Type 904, for 1964. This differed in many ways from the earlier car and was built for racing, its passenger seat there to comply with the rules rather than to be sat in. The 904 had its engine mounted ahead of the rear axle to create a mid-engined layout. The chassis had two sheet-steel, rectangular-section side members while the stylish coupe body was in fibreglass. It was planned to use a new flat-six engine under development for a new series but this was not available at first so the faithful four-cam unit in 2-litre form was further worked on for both power and reliability. It drove a five-speed gearbox, disc brakes were used and the driver's pedals could be adjusted over three locations as the seats were moulded into the floor pan.

Over 100 of these cars were built during early-1964 which enabled the 904 to be homologated. Early races showed where changes were needed but the car's potential was forcibly demonstrated when it won the 1964 Targa Florio with another second. Successes brought development and this brought more successes in 1965. The factory ran two cars with the flat-eight engine in 1964 and also the new flat-six which had a single overhead camshaft for each bank. About ten of these were built in 1965 but it was time for a new generation.

GRAND PRIX
YACHTING
BANCO

The Formula 1 Porsche fitted with the flat-eight engine based on the four-cam design, stretched out to more cylinders while retaining the 1500cc capacity.

Formula 1 Porsche in action: it won one Grand Prix, the French at Rouen in 1962, but did not run the following year.

ABOVE: Another view of the clean lines of the Formula 1 Porsche which included twin shrouds around the intakes to the two banks of cylinders. Years later the firm was again involved in Grand Prix racing but their real success was to remain in the sports car field.
BELOW: This 904 design took the firm on for 1964 and led to others in the same vein. The type was built in some numbers for sale and the works used one to win the 1964 Targa Florio.

THE 911 SERIES

Incredibly long-running, built for over 30 years and as popular now as when first launched late in 1964, the 911 series became the definitive Porsche in many guises. This one is from 1967.

During 1963 Porsche showed a new design at Frankfurt which became their longest-running model, even more so than the 356. Shown as the 901, it went into production late in 1964 as the 911 and is still listed.

The 911 followed Porsche traditions, was as fast as ever, but quieter, smoother and more comfortable. It retained the old engine location but the unit was an air-cooled, flat-six of 1991cc having a single, chain-driven overhead camshaft for each cylinder bank. This drove a five-speed gearbox which took the power to 15-inch wheels, which were stopped by disc brakes and all independently suspended. The body and chassis were one in steel, built by Reutter or Karmann as a coupe.

A year later, the Targa body was added. It had a detachable roof section ahead of the area above the occupants' heads and a folding rear window, changed in 1969 to a fixed assembly. During 1965 the 911 was joined by the 912 which was created to offer a relatively-lower priced Porsche by fitting the 1582cc flat-four pushrod engine into the 911 car and simplifying the interior trim. At first only four speeds were offered but by 1966 five were a choice, as was the Targa body style.

The flat-six engine which was first used in the 911. This one has fuel injection.

During 1966 the 911S was added to the line. This type had a tuned engine, better suspension, vented brake discs and alloy wheels. For 1967 an automatic transmission joined the option list, the unit retaining a clutch which was disengaged by a touch on the gear lever to give clutchless shifting of the four speeds. Listed as the Sportomatic it was altered to three speeds in 1975, by when the engine was larger, and was dropped in 1979, never having really found favour in the USA, for where it was developed.

Another variant, the 911R, was built in 1967 in small numbers. It was a competition model much reduced in weight and fitted with an engine up 60 bhp from standard. Later that year the 911L and 911T appeared, the first with improved trim and the second a reduced price thanks to some engine changes and a 912-type interior. In 1968 a Porsche 911 won the Monte Carlo Rally.

The original 911 was dropped for 1969 when the 911L was replaced by the 911E and all cars had an extended wheelbase. This improved both the weight distribution and the ride while the change came from placing the rear wheels further back. Both the E and S models changed to fuel injection in place of twin Webers.

In the second half of the 1960 decade Porsche continued their racing activity for they knew this would fine hone the cutting edge of their technology. From this came production competition cars, trimmed up more for road use and several special builds, many of which ultimately found their way into private hands.

In 1966 it was the Carrera 6, typed as the 906, which quickened the pulse. Built for racing, but sometimes seen on the road, the model used the flat-six engine from the 911, a tubular space frame and the 904 suspension. There was much use of magnesium alloy and titanium, the whole covered by a fibreglass body with full-wing doors. Around 50 were planned and the works ran at least one with the flat-eight engine.

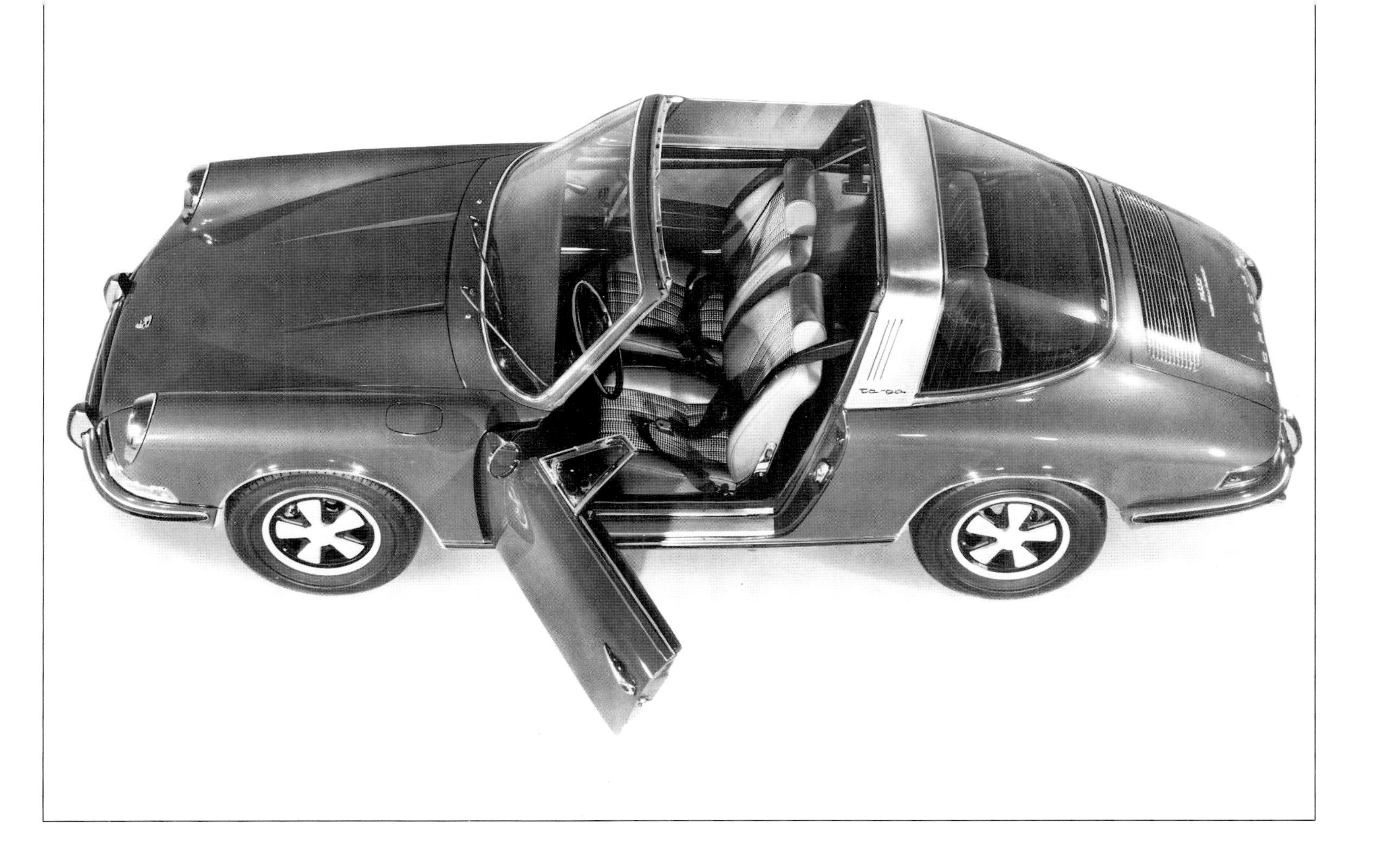

Targa body on the 911, with the detachable roof section removed, and a fixed rear window. The car offers a good choice between coupe and open-air motoring.

The following year saw a twin-cam version of the flat-six engine, and the 910 works car which was based on the 906. The 910 used both the six- and eight-cylinder engines, an uninspiring body, but had much improved handling. Ultimately some were sold off after a full works overhaul to help finance the future. The 907 which followed for later 1967 racing was a coupe aimed at Le Mans and had a right-hand drive which placed the driver on the inside of corners more often than not. It was fast, but unstable at high speed, although Porsche did learn how to deal with this in time. This led to the 908 which joined the 907 in 1968 and used a 3-litre version of the flat-eight engine, a six-speed gearbox and more body improvements. It was a successful race car. That year Porsche also built two hill climb cars as the type 909, short, open bodied and with the driver well forward. The 908 was used up to 1970 with coupe and open bodies but Porsche were planning their next move.

Early in 1969 the factory unveiled the 917 and by April they had 25 lined up to be homologated, just ten months after the project had begun. This made them eligible to run as 5-litre sports cars so the engine was a 4494cc flat-12, much as the Cisitalia designed back in 1947. By 1970 the 917 was developed, stretched to 4907cc and gave Porsche its first outright win at Le Mans after so many years of class and index results. Coming on top of finishing second in 1969, by a matter of yards, it was the start of Porsche dominance at the circuit.

There were major changes to the road models for 1970. The 911 models had the engine capacity raised to 2195cc, thanks to a larger bore, and continued in T, E and S forms in the coupe or Targa body styles. The 912 was replaced by two versions of the 914, a type that repeated the 904 mid-engined concept, but was built as a two-seater road car.

Above: Brochure picture of the 912 model which joined the 911 in 1965 to offer a lower-priced Porsche. This one is from 1968 and continued using the 1582cc flat-four pushrod engine rather than the ohc flat-six.
Below: Publicity shot of the 912 taken in 1968 which reflects the style of advertising and dress of the era.

In 1966 the 911S was added to the line to offer more power, suspension and brakes as well as alloy wheels. The name on the side ensured that all knew the marque name in the unlikely event of not recognising the shape.

ABOVE: During 1967 the 911L joined the range with improved trim. This is a 1968 model.
BELOW: Porsche 911S to the left and 911T. The latter had a 912-type interior to reduce the price.

Above: In 1966 the factory raced the 906. This one is seen competing on a dusty circuit where crowd control seems extremely limited.
Below: For the first part of 1967 the 910 was the works car. Later, some were sold off and this one is at a 1968 hill climb in private hands.

Above: The 910 was followed by the 907 for the later part of 1967. It was designed for Le Mans, hence the right-side seat to place the driver on the inside of most corners.
Below: Following the 907 came the 908 for 1968, fitted with the flat-eight engine and a six-speed gearbox.

ABOVE: Two decades later: a short-tail 908 out in Melbourne for a Sotherby's auction where its estimate was A$1.7 – $2 million (£800,000 – £1 million)!
BELOW: One of the 917 cars run at Le Mans in 1970, this one finishing second to its team mate. The psychedelic paint work led to it being called the 'hippie' car.

Flat-12 engine of the 917 on test at the works. Much of the basic design of this twin-cam unit came from the Cisitalia of 1947. Added to it was the fuel injection while the fan was mounted horizontally.

Stylish 911 Roadster with body by Bertone.

914 AND 916

The new 914 was the result of collaboration between Porsche and Volkswagen, the body built by Karmann in the Targa form with a detachable section above the seats. The mechanics were mainly as for the 911, with suspension, disc brakes and wheels all similar. Two engines were offered, the 914/4 using a 1679cc ohv flat-four unit from the VW441 car fitted with fuel injection in the USA. The 914/6 took the flat-six ohc engine from the 911 series fitted with carburettors and either engine sat just behind the seats and drove a five-speed gearbox. Two cars were built using the flat-eight engine, one being used by Ferry Porsche until the noise regulations caught up with it.

In 1972 the 911 series had the engine capacity increased once more, this time to 2341cc by lengthening the stroke. This was done more to cope with US emission laws than for power although this did go up, albeit on lower compression ratios. For the USA, all had fuel injection and four speeds, elsewhere they retained carburettors. The five speeds or the Sportomatic remained options, the suspension was improved and an air dam appeared under the front bumper of the 911S. An option for the others, it was soon standard for all.

The intended replacement for the 914/6 appeared as the 916 in 1972, much the same in line but with the top permanently fixed for extra body stiffness. The engine came from the 911S and there were five speeds. This light and fast-accelerating car was trimmed inside in leather and valour, colour keyed to the dash and panel – hardly in the Porsche image. Only twenty were built before it was decided not to take the car any further.

ABOVE: The 914 which replaced the 912 in 1970 to offer a mid-engined format powered by either a four-cylinder 1679cc VW or the 911 flat-six engine, the body in the Targa form and built by Karmann.
BELOW: A 914 in its 1974 form, by when only the four-cylinder VW engine was used, but enlarged to 1793cc. A 1971cc unit was an option, becoming the standard the next year.

A 911E as for 1974, this model replaced the 911L in 1969 and was offered in the coupe and Targa body styles.

Late in 1972 the famous Carrera name returned as the 911SC or Carrera RS. Destined for competition, the car had the body shell stripped down to the essentials to reduce its weight while the engine was bored out to 2687cc. A spoiler, called the 'ducktail', appeared on the rear of the car to hold it down.

Porsche planned to build 500 of these cars but customers queued to buy it and this run was sold by the end of the Paris show at which it made its debut. Part of this success was that the car was road-legal in Europe, even if not in the USA, and the firm offered an optional touring package with the interior of the 911S. The result was about 1,800 cars built by late-1973. From the Carrera RS the factory produced the RSR which was built for racing. Some 50 or so were produced during 1973, while the works continued the develop ment which included stretching the engine further to 2994cc, a far from easy task.

The 914/6 was dropped for 1973 when the 914/4 became available with a 1971cc engine as an option. Its standard engine was enlarged to 1793cc for 1974 but for 1975-76 the 2-litre one was fitted. The model was then dropped, never having been fully perceived as a Porsche, more as a VW. For all that, about 119,000 were built over the model's six-year life.

The base 911 model re-appeared in 1974 alongside the 911S, both of which now fitted the 2687cc engine to match the Carrera. The hardest task for the factory had been to meet the new bumper rules of the USA but Porsche managed this with a design that looked as if it had always been there. The Carrera was offered with the Targa body as well as the coupe while the RS and RSR were built for competition and fitted with the 2994cc engine. However, as in 1973, a good number finished up being used on the road.

A Porsche photograph: the car carrying a code of FLA which breaks the tradition of using numbers. A concept design, but one that retained the rear engine mounting, possibly meant for Volkswagen.

In 1975 Porsche unveiled yet more excitement, the 911 Carrera Turbo. This took the concept a big step forward for the car was fitted with the 2994cc engine as well as the turbocharger. A mock-up had been shown at Frankfurt late in 1973 but a year later in Paris it was the real thing. The body was from the Carrera, complete with ducktail, the transmission was strengthened and the car came very fully equipped. It was a supercar, very fast, very expensive and with stunning lines. In the USA it was listed as the Turbo Carrera and only the coupe body was offered.

Porsche listed a Silver Anniversary 911S in 1975. Each had a plate carrying Ferry's signature and a serial number. All four versions of the 911 had much improved rust resistance for 1976, due to the use of zinc-coated steel, in production. For that year the Carrera fitted the 2994cc engine but was only sold in Europe as was the 911. The USA had the 911S and the Turbo Carrera, still sold as the Turbo in Europe.

The works 917 won Le Mans for a second time in 1971 and the series was most successful at other venues including the Can-Am series. Some of its technology went into the Carrera RSR cars and from them came the Turbo RSR with a 3-litre, flat-six engine. Around 30 were built in 1976 and this led on to the works 935 which had a 2.8-litre turbo engine, about ten being built for private use in 1977. Finally there came the 936 which had its 2.1-litre engine set ahead of the rear axle in a tubular space frame covered by an open body. The 936 won at Le Mans in 1976 and 1977, also at many other circuits, while a private 935 won at Le Mans in 1979.

The 912 came back in 1976 to fill a gap lower in the price range. Aimed at the US market, it used the 1971cc VW flat-four engine, a five-speed gearbox and the 911 coupe or Targa body with a simplified trim. Listed as the 912E, it was only sold for one year, in order to hold the fort until a new model was ready.

Above: Late in 1972 a revered name was revived when the 911SC appeared as the Carrera RS.
Below: The 1973 version of the 914 with detachable roof section in place.

The Carrera RS model showing the rear spoiler or ducktail.

The front bumper that Porsche came up with to meet the stringent US rules that came into effect for 1974, in this case on a 1975 911 Turbo.

OPPOSITE: The 911E as for 1973, in this case the coupe body has a sun roof.

TYY 39L

Another view of a 914 with its top in place, a model that was never quite perceived as a true Porsche, more a VW, but nevertheless still sold well.

The 911 Carrera Turbo which joined the range for 1975 to offer much more performance from its 2994cc engine – fast, expensive and a supercar with super looks.

Above: Rear view of the 1975 911 Carrera Turbo which had the ducktail fitted to ensure the rear wheels stayed well in contact with the road.
Below: Standard 1975 911 coupe which managed without the rear spoiler but retained the fine Porsche line and style.

ABOVE: The 911S of 1975 in Targa body form complete with ducktail.
BELOW: In 1976 the 911 Carerra was fitted with the 2994cc engine but only sold in Europe.

924 AND 928

The Group 5 type 935 which extended the power limits even further, developing over 600 bhp on occasion.

In one sense Porsche went back to their 356 origins with the 924, using VW parts and assemblies for much of the car. On the other hand, they changed their history, building a car with both water-cooling and a front-mounted engine. However, the 356 was conceived as a Porsche built using modified VW parts while the 924 was meant to be a VW but became a Porsche.

Company changes and problems affected the issue but the outcome went on line late in 1975, selling at first in Europe and then in the USA. The resultant car introduced a new body style with pop-up headlamps and one of the lowest drag coefficients in the car world of that time.

Under the bonnet went a four-cylinder, in-line, 1984cc engine, designed by VW and built by Audi. It was laid over by 40 degrees to ensure a low bonnet line and had a single overhead camshaft driven by a toothed belt. There was fuel injection, and a special, finned sump, angled to suit the installation, was fitted to offset the lack of an oil cooler.

The clutch went behind the engine but at this point the Porsche individuality came back for the four-speed gearbox went at the back of the car with the transaxle as usual. This moved some weight onto the rear wheels and made the car easier to handle when driven hard. Many of the suspension parts were from VW cars as were the brakes, which were discs at the front but drum rear.

Very different in many ways, the 924 models used a water-cooled, front-mounted VW engine laid over to reduce the bonnet height.

In 1977 a three-speed automatic transmission became available for the 924 which had found favour compared to the 914. Later that year a five-speed gearbox was added, much desired to fill the jump from second to third of the original. A Martini Edition 924 was built in 1977, followed by a Sebring '79 for the US market. The first was in white, the second, about 1300 cars, in bright orange-red. Yet another special edition, the Le Mans, was offered in Europe in 1980. Meanwhile, the four versions of the 911 ran on and a further new model was launched, the 928.

The idea behind the 928 came before the 924 and the car was first seen in public early in 1977 at the Geneva show. Conceived as a grand tourer, it was intended to be at the top of the range, as fully equipped and as luxurious as was feasible in the 2+2 body shell, and to have the same long production life as the 356 and 911. It copied the 924 in having a front-mounted, water-cooled engine and the gearbox in unit with the rear axle, but differed in many ways.

The Le Mans edition of the 924 which was listed for Europe in 1980, showing the Porsche lines.

The engine was an all-alloy, V-8 of 4474cc with a single overhead camshaft driven by a toothed belt for each bank. It drove either a five-speed gearbox or a three-speed automatic transmission and the chassis followed the Porsche convention of independent suspension and disc brakes for all wheels.

Zinc-coated steel was used for the body shell but the doors and engine hatch were in aluminium alloy. Inside went all the essentials of GT motoring such as air conditioning, cruise control, electric windows and much more, while the pop-up headlamps were used. The result was as planned, a grand touring Porsche which behaved as such cars should, so that newcomers to the marque were happy. Old Porsche hands were less pleased as the feel differed, but only they realised this.

The 911 range ran on while the 924 and 928 were launched but had their turn for change in 1978. The Carrera became the 911SC while the 911 Turbo increased in size to 3299cc. For 1979 these continued while the other ranges had additions in the form of the 924 Turbo and the 4664cc 928S, both quicker than the standard cars. Quicker still was the 924 Carrera GT of which about 400 were built for competition but, as usual, some became road-legal in Europe. It was just the same with the Carrera GTS racer of 1980 while the 911SC Weissach Coupe joined the range as a limited (400) edition which had the looks of the Turbo but without its engine. There was also a 924 Weissach built for North America in 1981, followed by a 928 Weissach finished in a metallic gold and complete with a luggage set in matching leather.

ABOVE: By 1979, when this 924 was built, there were options of automatic transmission or five speeds in place of the stock four.
BELOW: As a contrast there was this Indianapolis car for Hawaiian Danny Ongais to drive in 1980 – a 'brick-yard' special.

ABOVE: Fine 1977 coupe 911 Carrera with right-hand drive for the UK market.
BELOW: Conceived as a grand tourer, the 928 was launched in 1977 and achieved its purpose using a front-mounted, watercooled V-8 engine of 4474cc. Quick and comfortable.

ABOVE: Graceful lines of the 928 which included the pop-up headlamps.
BELOW: For 1978 the Carrera became the 911SC. This is the 1979 model complete with ducktail.

ABOVE: The 1979 edition of the 911 Turbo which had its engine enlarged to 3299cc for the previous year.
BELOW: Fine overhead shot of the 911 Turbo which displays the ducktail to advantage.

ABOVE: Front overhead view of the 911 Turbo to show the sleek lines of this ever-popular model.

Above: A 911 Turbo Martini for 1981, a period when a number of special editions were offered.

Opposite: Rear view of the 911 Turbo Martini with the hatch up. Engine cooling in the small bay was a problem that Porsche worked hard to solve.

PORSCHE
AAH 379X

944

Fine 1982 911SC which continued the body style of the series, changed but not really altered since 1964.

The 911, 924 and 928 cars ran on into the early-1980s, subject to the continuous development characteristic of Porsche, and were joined by a new model in 1982. This was the 944 which became the car that the 924 should have been. The concept of a water-cooled, in-line, four-cylinder engine driving a rear-mounted transmission remained but the result differed.

The 944 engine displaced 2479cc and had a single overhead camshaft driven by a toothed belt. To counter the inherent roughness of the four there were two balance shafts within the crankcase, belt driven in opposite directions to each other at twice engine speed. These generated opposing forces to those of the engine so that all balanced in the crankcase and did not disturb the outer world. The result was a very smooth car, faster than the 924 and well equipped. Its relatively low price ensured its success in the salesroom.

In 1982 the 944 joined the range to continue the concept of the 924 and 928 but with a very smooth four-cylinder 2479cc engine.

Meanwhile, the works began a run of wins at Le Mans, the first in 1981 with a modified 936 car. In 1982 there was the 956 having a 2.6-litre, flat-six engine with twin turbos, which by 1984 had become the 962. The engine gradually increased in size, became water-cooled and Porsche won Le Mans each year from 1981 to 1987. They also succeeded in the Paris-Dakar Rally in 1984 with a four-wheel drive 911 Carrera and in 1986 with a 959.

In 1983 the 911 range was joined by the 911SC Cabriolet, the first such from Porsche for nearly twenty years. That year the US market had the 928S in a 4644cc capacity while the basic 928 was dropped. Changes came for 1984 when the 911SC became the 911 Carrera to revive a famous name and fit a larger 3164cc engine. The 924 Turbo, never too popular, was dropped while a fantastic Porsche to come continued to tantalise enthusiasts. First seen at a Frankfurt show, this was the Gruppe B race car which, in time, evolved into the 959.

There was one important change in 1985 when the 928S had new cylinder heads and an increase in capacity to 4957cc. The changes introduced not only four-valve heads but also twin overhead camshafts to create an even better great car. In 1986 there was more, for the 924S was added, using the 2479cc engine and much else from the 944 while that range extended to three models. One was the 944 Turbo which kept the existing engine with less compression

This was the 956 which won at Le Mans in 1985, the firm's fifth victory in a row and tenth since the first in 1970.

but with the blower, the other the 944S. This took some of the 928S technology, even if it used none of the detailed parts, to create another four-valve, twin overhead camshaft engine.

The range ran on for 1987 but that year saw the fabulous 959 out on the roads of Europe. This was an enormously-expensive road car that could match most racing cars. It was powered by a flat-six, 2850cc engine that combined air and water cooling, had four valves per cylinder, twin turbochargers, fuel injection and pushed out some 450 bhp. This drove via a six-speed gearbox to all four wheels, the chassis represented the pinnacle of Porsche achievement and the top speed was given as over 190 but under 200 mph. It was not offered in the USA, not due to the speed limits but because the emission and bumper rules meant changes not warranted by the numbers to be built, 200 in total.

Fine photograph of a 928S taken in 1984, the car blending well into the scene.

In time the 956 moved on to become the 962; this one is seen in 1989. It used a 3-litre, flat-six, water-cooled engine fitted with twin turbochargers.

In 1988 there was a 944 Turbo S, a short production run for that year only but a rather special car. There was also the 911 Club Sport, a simplified Carrera that lost both weight and cost by removing many items often thought of as extras. There were some engine and chassis changes as well to sharpen the car for racing at club level, a hint of the Porsche past. In addition there was a short production run of a Silver Anniversary 911 in the Coupe, Cabriolet and Targa bodies.

In 1989 the 911 Carrera was replaced by the Carrera 4 which had four-wheel drive, a new 3600cc flat-six engine, and a new floor pan plus suspension. The drive to the front was a simple extension forward and then out to the wheels, all four being continuously driven via differentials. This was based on the system used in the 959, but improved and simplified to give traction control, a further safety item to add to the ABS brakes. A neat feature was the rear spoiler which was hinged to lay flat to the body until the car was up to around 50 mph. It then raised automatically but folded again at 6 mph.

The 911 Turbo and Club sport continued for 1989 and were joined by the 911 Speedster, an 800-off run of cars based on the Carrera Cabriolet but with a modified body. The 928S stayed as it was but the 944 was stretched to 2671cc; the twin-cam version became the 944S2, was enlarged to 2990cc and became available in Cabriolet form for the first time. The 924 cars were axed from the range.

By 1983 there was a 911SC Cabriolet listed. This is its form for the following year, with the top raised.

A Carrera 2 joined the 911 range for 1990, in essence the Carrera 4 without the front-wheel drive. However, it could be fitted with Tiptronic transmission which gave the driver the choice of a normal automatic or a manual, but clutchless, four speeds. To achieve this, the gear lever had a double gate, the left slot offering the usual auto markings, the right side a simple forward for up, back for down selection.

The Carrera 4 and 911 Turbo remained listed, as did the 944S and 944 Turbo, while the 928S changed its code to 928. All continued for 1991 when the 911 Turbo had yet another face lift to continue its long life. The engine was worked on to push its power up to 320 bhp and the chassis improved to bring its abilities more in line with those of the Carrera models.

Above: The same 911SC Cabriolet, but with the top lowered for some fresh air motoring.

Opposite: The 1984 model of the 928S as seen in many other drivers' mirrors, coming up fast and quiet from astern, showing its tidy lines.

THE 928S

Right: Cockpit of the 928S in 1984, in this case fitted with the automatic transmission option. Clear dials right in front of the driver and minor controls at fingertip.

Below: The 911 Carrera Cabriolet with the top up for 1988, a time when the upwardly mobile favoured the marque.

LEFT: Power unit of the 911 Carrera, dominated by the massive cooling fan. Under the covers lies the flat-six engine and some advanced technology.

BELOW: This is the Turbo version of the 924 which failed to find favour with buyers so was dropped in 1984.

For 1985 the 928S changed to twin overhead camshafts and four-valve heads for its engine which increased in size to 4957cc. This is the 1987 model.

LEFT: Engine bay of the 928S of 1987 which gives a very good idea as to how tight all the machinery had to be packed in, along with the ancillary items demanded for such a car.

BELOW: Seating and luggage area of the 928S, fine for two but really rather cramped for four.

Above: The 924S was added for 1986 and used the 2479cc engine and other features from the 944 while keeping its own body style.

Right: Cabin area of the 924S which shared some features with the other cars.

ABOVE: Another newcomer for 1986 was the 944 Turbo which was much like the standard model but had the blower as well.

LEFT: Interior of the 944 Turbo which differed in many respects from the 924 and 928 to create another individual style.

ABOVE: This is the fabulous Porsche 959, an incredibly expensive car that used a very special version of the flat-six engine with twin turbochargers, fuel injection and four valves per cylinder to push out 450 bhp.
BELOW: Rear view of the 959 which had a six-speed gearbox, four-wheel drive, the best of chassis, and a 190 mph top speed.

ABOVE: Porsche development represented from the right by a 2-litre 911 coupe from 1965, a 3-litre 911 Turbo from 1975 and the fantastic 959 of 1988 but all from the same family.
BELOW: The 911 Turbo S built in 1988 and fitted with enough extras to make it a special car while retaining the series mould.

ABOVE: Further view of the 911 Turbo S which came with the rear spoiler and extra vents in the body sides just ahead of the rear wheels.
BELOW: The 911 Carrera Cabriolet as for 1988, with its hood down for a summer drive.

Fine frontal view of a standard 911 from 1989, once again showing how advanced the lines and style were when the car was first seen back in 1964.

Stock 944 of 1989, with its headlamps concealed under the pop-up covers.

Above: The 911 SE Cabriolet of 1989, its message saying 'come and enjoy me'.
Below: The 928S kept its headlamps exposed even when not raised and continued with its fine lines and high performance.

ABOVE: By 1989 the 944S, using a twin overhead camshaft, four-valve engine, had increased in size to 2990cc and became the 944S2.
BELOW: For 1989 the 944S2 also became available in Cabriolet form for the first time to offer their buyers the option of fresh air motoring.

ABOVE: The control lever for the Tiptronic transmission system which gave the choice of automatic or manual gears.

ABOVE: Tiptronic gear and mode indicator incorporated into the rev-counter.

A 944 Turbo in its 1990 form and one of the listed colours. This was to be its penultimate production year.

Grand touring remained the strong point of the 928. This 1990 version is as fully equipped as ever and fast and comfortable so able to cover much ground quickly.

968

Late in 1991 Porsche introduced the 944 replacement as the 968. It utilised elements from many of their cars combined with further technology so that much was new. The engine was the 2990cc four from the 944, complete with balancer shafts and twin overhead camshafts for the four valves per cylinder. New was their Vario-Cam timing system which allowed the timing to alter to meet the demand. Despite the increase in torque this gave, the transmission was by six-speed gearbox with the option of the four-speed Tiptronic. Much of the chassis was based on that of the 944 and various option packs were listed as usual, the car available with either coupe or cabriolet body.

The 968 Cabriolet, the series which replaced the 944 and used technology from several of the Porsche series. The engine had the Vario-Cam valve timing system.

In 1992 the 911 Carrera RS returned along with RSL Lightweight and Touring versions to continue the Carrera theme of more performance from less weight. This resulted in a performance from the RSL akin to that of the 911 Turbo but with more noise and a different feel on the road, aimed as the car was to circuit use. That year saw the revival of the American Roadster as a 3.6-litre 911 cabriolet and the run down of the 944 models.

Details of the Vario-Cam valve system as shown in the brochure for buyers to study.

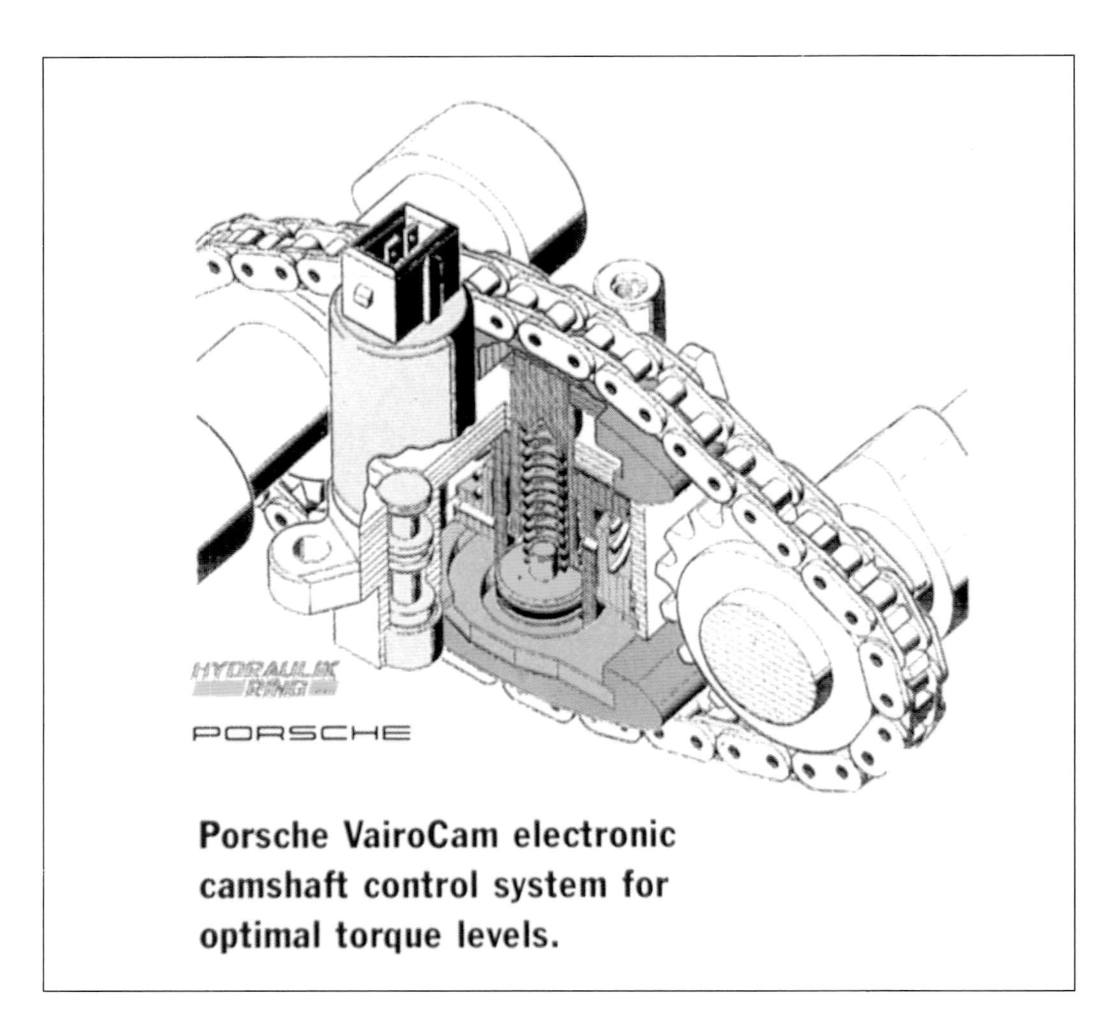

In 1993 the 968 Club Sport joined the standard 968 and lost weight by removing many of the electrically-operated items, such as windows, and a number of other features. The result was a crisper car. In similar vein was the 911 RS America which lost weight and had a fixed rear spoiler to lift the performance. The 928 became the 928GTS and stretched out to 5397cc while the other models ran on.

The 1994 911 Carrera had a new rear suspension system along with many other changes including the six-speed gearbcx. Also new for 1994 was the 968 Sport which took the Club Sport and put back some of the deleted features in response to customer demand.

For 1995 a new, much revised, 911 Carrera 4 was introduced. Its four-wheel drive was extensively altered, much lighter than before, and gave the car as good or better performance than the stock version. Both coupe and cabriolet models were offered, the new drive system enhancing the handling under hard cornering.

The automatic 911 Carrera had a further refinement added to its Tiptronic system for 1995 in the form of steering wheel mounted rocker switches for gear selection, allied to an intelligent shift program. This monitored sensors to ensure the optimum ratio for all conditions.

Thus, the various Porsche models continued to give pleasure to the driver, for that is what the firm is all about – performance, style, pleasure and advance.

ABOVE: Four-wheel drive Carrera 4 from 1992, a model introduced in 1989 which used the 3600cc flat-six engine and much else that was new.
BELOW: This is the 968CS or Club Sport for 1993, a lightweight version built for competition.

ABOVE: The 1994 version of the 911 Carrera remained much in the series mould.
BELOW: Fabulous! The 911 Carrera 4 for 1995 gave the driver a highly advanced four-wheel drive system for traction, handling and control.

LEFT: Revised for 1995, the Tiptronic automatic used wheel-mounted switches for gear changing, as found in Formula 1 racing cars.

BELOW: This is the 911 Carrera Cabriolet with the hood raised.

Above: The 911 Carrera Cabriolet, hood down and open for the fresh air.
Below: A look back at Porsche history with three of their competition cars which finished up as road-legal vehicles for some delighted owner. Very early to quite late.

PORSCHE MODELS

MODEL	YEAR	CC	ENGINE	BODY
356	1948	1131	F-4 ohv	open
356/2 1100	1949-54	1086	F-4 ohv	coupe, cabriolet
356/2 1100	1949-50	1131	F-4 ohv	coupe, cabriolet
356/2 1300	1951-55	1287	F-4 ohv	coupe, cabriolet
356/2 1300A	1954	1287	F-4 ohv	coupe, cabriolet
356/2 1300S	1955	1287	F-4 ohv	coupe, cabriolet
356/2 1500	1952-55	1488	F-4 ohv	coupe, cabriolet
356/2 1500	1952-55	1488	F-4 ohv	America, Speedster
356/2 1500S	1953-55	1488	F-4 ohv	coupe, cabriolet
356A 1300, 1300S	1956-57	1290	F-4 ohv	coupe, cabriolet, Speedster
356A 1600, 1600S	1956-59	1582	F-4 ohv	coupe, cabriolet, Speedster
356B 1600, 1600S	1960-63	1582	F-4 ohv	coupe, cabriolet, Roadster
356B 1600S-90	1960-63	1582	F-4 ohv	coupe, cabriolet, Roadster
356C 1600C, 1600SC	1963-65	1582	F-4 ohv	coupe, cabriolet
911	1965-68	1991	F-6 ohc	coupe, Targa
911S	1966-69	1991	F-6 ohc	coupe, Targa
911R	1967	1991	F-6 ohc	coupe
911L	1967-68	1991	F-6 ohc	coupe, Targa
911T	1967-69	1991	F-6 ohc	coupe, Targa
911E	1969	1991	F-6 ohc	coupe, Targa
911S, 911T, 911E	1970-71	2195	F-6 ohc	coupe, Targa
911S, 911T, 911E	1972-73	2341	F-6 ohc	coupe, Targa
911SC Carrera RS	1973	2687	F-6 ohc	coupe
911 Carrera RSR	1973-75	2994	F-6 ohc	coupe
911, 911S	1974-77	2687	F-6 ohc	coupe, Targa
911 Carrera	1974-75	2687	F-6 ohc	coupe, Targa
911 Carrera RS	1974-75	2994	F-6 ohc	coupe
911 Carrera Turbo	1975-77	2994	F-6 ohc	coupe
911 Carrera	1976-77	2994	F-6 ohc	coupe, Targa
911SC	1978-83	2994	F-6 ohc	coupe, Targa, cabriolet
911 Turbo	1978-95	3299	F-6 ohc	coupe, Targa, cabriolet
911 Carrera	1984-88	3164	F-6 ohc	coupe, Targa, cabriolet
911 Club Sport	1988-89	3164	F-6 ohc	coupe
911 Speedster	1989	3164	F-6 ohc	Speedster
911 Carrera 4	1989-95	3600	F-6 ohc	coupe, Targa, cabriolet
911 Carrera 2	1990-93	3600	F-6 ohc	coupe, Targa, cabriolet
911 America	1992	3600	F-6 ohc	Roadster
911 RS Carrera	1992	3600	F-6 ohc	coupe
911 RSL Carrera	1992	3600	F-6 ohc	coupe
911 RS America	1993	3600	F-6 ohc	coupe
911 Carrera	1994-95	3600	F-6 ohc	coupe, cabriolet
912	1965-69	1582	F-4 ohv	coupe, Targa
912E	1976	1971	F-4 ohv	coupe
914/6	1969-72	1991	F-6 ohc	Targa
914/4	1969-73	1679	F-4 ohv	Targa
914/4	1973-76	1971	F-4 ohv	Targa
914/4	1974	1793	F-4 ohv	Targa

MODEL	YEAR	CC	ENGINE	BODY
916	1972	2341	F-6 ohc	coupe
924	1976-88	1984	I-4 ohc	coupe
924 Turbo	1979-83	1984	I-4 ohc	coupe
924 Carrera GT	1981	1984	I-4 ohc	coupe
924 Carrera GTS	1981	1984	I-4 ohc	coupe
924S	1986-88	2479	I-4 ohc	coupe
928	1977-82	4474	V-8 ohc	coupe
928S	1979-84	4664	V-8 ohc	coupe
928S	1983-84	4644	V-8 ohc	coupe
928S	1985-89	4957	V-8 dohc	coupe
928	1990-92	4957	V-8 dohc	coupe
928 GTS	1993-95	5397	V-8 dohc	coupe
944	1982-88	2479	I-4 ohc	coupe
944 Turbo	1986-92	2479	I-4 ohc	coupe
944S	1986-88	2479	I-4 dohc	coupe
944 Turbo S	1988	2479	I-4 ohc	coupe
944	1989	2671	I-4 ohc	coupe
944 S2	1989-92	2990	I-4 dohc	coupe, cabriolet
959	1987-88	2849	F-6 dohc	coupe, cabriolet
968	1992-95	2990	I-4 dohc	coupe
968CS	1993-95	2990	I-4 dohc	coupe
968S	1994-95	2990	I-4 dohc	coupe

PRODUCTION SPORTS-RACING AND WORKS CARS

MODEL	YEAR	CC	ENGINE	BODY
Glöckler	1951	1500	F-4 ohv	open, coupe
550	1953	1488	F-4 ohv	open, coupe
550/1500RS	1954-55	1498	F-4 dohc	open
Carrera 1500GS	1956-57	1498	F-4 dohc	coupe, cabriolet, Speedster
550A/1500RS	1956-59	1498	F-4 dohc	open, coupe
Carrera 1500GS GT	1957-59	1498	F-4 dohc	coupe, Speedster
Carrera 1500GS DL	1957-59	1498	F-4 dohc	coupe, cabriolet, Speedster
Carrera RSK	1957-58	1498	F-4 dohc	open
Carrera RSK	1957-58	1587	F-4 dohc	open
Carrera 1600GS GT	1958-59	1587	F-4 dohc	coupe, Speedster
Carrera 1600GS DL	1958-59	1587	F-4 dohc	coupe, cabriolet, Speedster
Carrera 1600GS	1959-60	1587	F-4 dohc	coupe
1500 RSK	1959	1498	F-4 dohc	open
Abarth Carrera GTL	1960-61	1587	F-4 dohc	open, coupe
Carrera 2000GS	1962-65	1968	F-4 dohc	coupe
904 Carrera GTS	1964	1968	F-4 dohc	open, coupe
904 Carrera GTS	1965	1991	F-6 ohc	open, coupe
906 Carrera 6	1966	1991	F-6 ohc	coupe
910	1967	1991	F-6 ohc	coupe
907	1968	1991	F-6 ohc	coupe
908	1969	2996	F-8 dohc	open, coupe
917	1969	4494	F-12 dohc	coupe
934 Turbo RSR	1976	2994	F-6 ohc	coupe
935	1977	2808	F-6 ohc	coupe